The Crab nebula, M1, NGC 1952 (Taurus A) is a supernova remnant.

All times shown are in Universal Time (UT), which is the same as Greenwich Mean Time (GMT). Therefore, the reader should adjust the Universal Time to his/her time zone, which may change the dates of certain lunar events. Astronomical data is from the US Naval Observatory and the NASA/Goddard Space Flight Center.

BrownTrout Publishers attempts to record event dates accurately to the full extent permissible by applicable law. BrownTrout Publishers disclaims all warranties, express or implied, including, without limitation, implied warranties of merchantability and fitness for a particular purpose. Events can differ from dates shown due to regional or sectarian observances. BrownTrout shall not be responsible or liable for any reliance on the displayed event dates.

USA

BrownTrout Publishers
P.O. Box 280070
San Francisco, CA 94128-0070 USA
Toll Free: 800 777 7812
(1) 310 607 9010
Fax: 310 607 9011
sales@browntrout.com

UNITED KINGDOM

BrownTrout Publishers Ltd.
P.O. Box 201
Bristol BS99 5ZE England UK
(44) 117 317 1880
UK Freephone: 0800 169 3718
sales@browntroutuk.com

AUSTRALIA & NEW ZEALAND

BrownTrout Publishers Pty. Ltd.
12 Mareno Road
Tullamarine VIC 3043, Australia
03 9338 4766
Outside Australia: (61) 3 9338 4766
Australia Toll Free: 1 800 111 882
New Zealand Toll Free: 0800 888 112
sales@browntrout.com.au

CANADA

BrownTrout Publishers Ltd.
55 Cork Street East
Unit 206
Guelph ON N1H 2W7, Canada
(1) 519 821 8882
Canada Toll Free: 1 888 254 5842
Fax: (1) 519 821 1012
sales@browntrout.ca

MEXICO

Editorial SalmoTruti, SA de CV
Hegel 153 Int. 903, Colonia Polanco,
Del. Miguel Hidalgo,
11560 Mexico D.F., Mexico
(52-55) 5545 0492
Mexico Toll Free: 01 800 716 7420
ventas@salmotruti.com.mx

EARTH-FRIENDLY

BrownTrout products are printed on Forestry Stewardship Council (FSC) certified paper from managed fores
and are printed with soy and vegetable-based inks, which are less harmful to the
environment than petroleum-based alternatives.

For more information about BrownTrout publications, please visit our website
www.browntrout.com

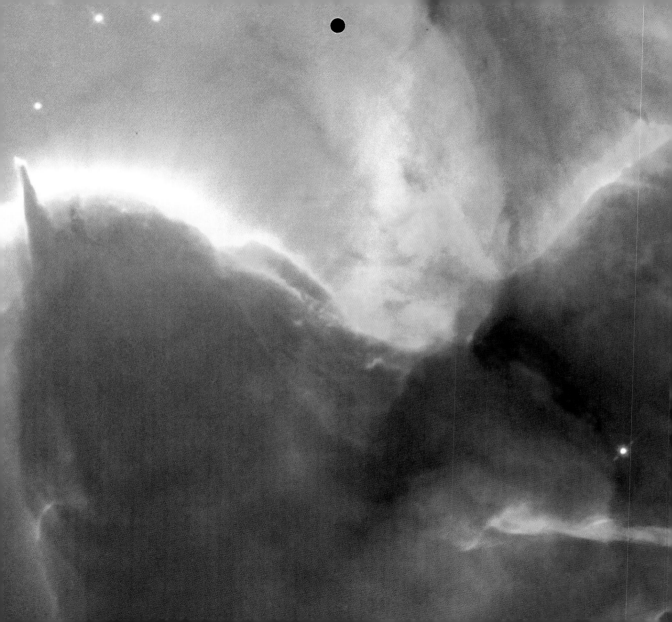

JANUARY 2011

JANVIER · ENERO · JANUAR

SUNDAY dim·dom·son	MONDAY lun·lun·mon	TUESDAY mar·mar·die	WEDNESDAY mer·mièr·mit	THURSDAY jeu·jue·don	FRIDAY ven·vier·fre	SATURDAY sam·sáb·sam
						1 New Year's Day Jour de l'An Año Nuevo Neujahr Kwanzaa ends
2	**3** Perihelion 19:00 U.T. Day after New Year's Day (NZ) Bank Holiday (UK)	**4** New Moon 9:03 U.T. ● Solar Eclipse (Partial) 8:51 U.T. Bank Holiday (SCT)	**5**	**6** Epiphany Épiphanie Día de los Reyes Heilige Drei Könige	**7**	**8**
9	**10**	**11**	**12** First Quarter 11:31 U.T. ◑	**13**	**14**	**15**
16	**17** Rev. Martin Luther King, Jr. Day (US)	**18**	**19** Full Moon 21:21 U.T. ○	**20**	**21**	**22**
23	**24**	**25** Burns Night (SCT)	**26** Last Quarter 12:57 U.T. ◑ Australia Day (AU)	**27** Holocaust Memorial Day (UN)	**28**	**29**
30	**31**	For more calendar events, visit www.browntrout.com				

DECEMBER 2010

28	29	30	**1**	2	3	4
5	6	7	8	9	10	11
12	13	14	15	16	17	18
19	20	21	22	23	24	25
26	27	28	29	30	31	1
2	3	4	5	6	7	8

FEBRUARY 2011

30	31	**1**	2	3	4	5
6	7	8	9	10	11	12
13	14	15	16	17	18	19
20	21	22	23	24	25	26
27	28	1	2	3	4	5
6	7	8	9	10	11	12

Astronomy

Simultaneous eclipse of three of Jupiter's moons

SUNDAY dim·dom·son	MONDAY lun·lun·mon	TUESDAY mar·mar·die	WEDNESDAY mer·miér·mit	THURSDAY jeu·jue·don	FRIDAY ven·vier·fre	SATURDAY sam·sob·sam
30	31	1	2 Groundhog Day Día de la Candelaria (MX)	3 New Moon 2:31 U.T. ● Chinese New Year – Year of the Rabbit	4	5 Día de la Constitución (MX)
6 Waitangi Day (NZ)	7	8	9	10	11 First Quarter 7:18 U.T. ◑	12 Lincoln's Birthday (US)
13	14 Valentine's Day Saint-Valentin Valentinstag Día del Amor y la Amistad (MX)	15	16	17	18 Full Moon 8:36 U.T. ○	19
20	21 Presidents' Day (US) Family Day (AB, ON, SK - CAN)	22 Washington's Birthday (US)	23	24 Last Quarter 23:26 U.T. ◑ Día de la Bandera (MX)	25	26
27	28		For more calendar events, visit www.browntrout.com			

JANUARY 2011

S	M	T	W	T	F	S
26	27	28	29	30	31	1
2	3	4	5	6	7	8
9	10	11	12	13	14	15
16	17	18	19	20	21	22
23	24	25	26	27	28	29
30	31	1	2	3	4	5

MARCH 2011

S	M	T	W	T	F	S
27	28	1	2	3	4	5
6	7	8	9	10	11	12
13	14	15	16	17	18	19
20	21	22	23	24	25	26
27	28	29	30	31	1	2
3	4	5	6	7	8	9

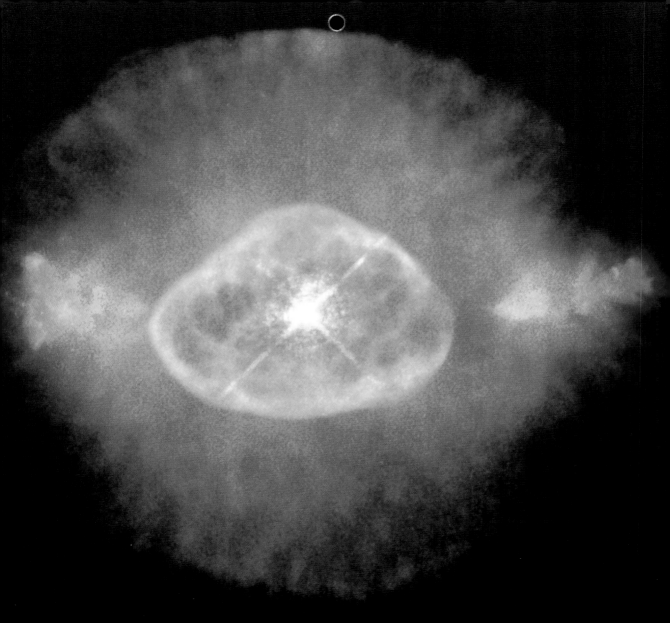

MARCH 2011
MARS · MARZO · MÄRZ

SUNDAY dim·dom·son	MONDAY lun·lun·mon	TUESDAY mar·mar·die	WEDNESDAY mer·miér·mit	THURSDAY jeu·jue·don	FRIDAY ven·vier·fre	SATURDAY sam·sáb·sam
		1 St. David's Day (WAL)	**2**	**3**	**4** New Moon 20:46 U.T. ●	**5**
6	**7** Labour Day (WA - AU) Great Lent begins (Orthodox)	**8** Shrove Tuesday Fat Tuesday Mardi gras Martes de Carnaval Fastnacht International Women's Day	**9** Ash Wednesday Mercredi des Cendres Miércoles de Ceniza Aschermittwoch	**10**	**11**	**12** First Quarter 23:45 U.T. ◐
13 Daylight Saving Time begins (US; CAN)	**14** Commonwealth Day Labour Day (VIC - AU) Eight Hours Day (TAS - AU)	**15**	**16**	**17** St. Patrick's Day Saint-Patrick San Patricio	**18**	**19** Full Moon 18:10 U.T. ○
20 Vernal Equinox 23:21 U.T. Journée internationale de la Francophonie Int'l Speakers of French Day	**21** Natalicio de Benito Juárez (MX)	**22**	**23**	**24**	**25**	**26** Last Quarter 12:07 U.T. ◐
27 European Union Daylight Saving Time begin	**28**	**29**	**30**	**31**	For more calendar events, visit www.browntrout.com	

Astronomy

Supernova remnant E0102 in the Small Magellanic Cloud

SUNDAY dim·dom·son	MONDAY lun·lun·mon	TUESDAY mar·mar·die	WEDNESDAY mer·mièr·mit	THURSDAY jeu·jue·don	FRIDAY ven·vier·fre	SATURDAY sam·sáb·sam
					1 April Fools' Day	**2**
3 Mothering Sunday (UK) Daylight Saving Time ends (NZ)	**4** New Moon 14:32 U.T. ●	**5**	**6**	**7**	**8**	**9**
10	**11** First Quarter 12:05 U.T. ◐	**12**	**13**	**14**	**15**	**16**
17 Palm Sunday Dimanche des Rameaux Domingo de Ramos	**18** Full Moon 2:44 U.T. ○ Passover begins at sundown	**19**	**20**	**21** Maundy Thursday Jeudi saint Jueves Santo Birthday of Queen Elizabeth II	**22** Earth Day Good Friday Vendredi saint Viernes Santo Karfreitag Bank Holiday (UK)	**23** Holy Saturday Samedi saint Sábado de Gloria St. George's Day (ENG)
24 Easter Sunday Pâques Domingo de Pascua Ostersonntag Pascha (Orthodox)	**25** Last Quarter 2:47 U.T. ◑ Holy Monday Bank Holiday (ENG, WAL, NIR) ANZAC Day (AU; NZ)	**26**	**27**	**28**	**29** Arbor Day (US)	**30** Día del Niño (MX) Koninginnedag (NL)

For more calendar events, visit www.browntrout.com

MARCH 2011						
27 28	1	2	3	4	5
6	7	8	9	10	11	12
13	14	15	16	17	18	19
20	21	22	23	24	25	26
27	28	29	30	31	1	2
3	4	5	6	7	8	9

MAY 2011						
1	2	3	4	5	6	7
8	9	10	11	12	13	14
15	16	17	18	19	20	21
22	23	24	25	26	27	28
29	30	31	1	2	3	4
5	6	7	8	9	10	11

Astronomy

MAY 2011
MAI · MAYO · MAI

SUNDAY dim·dom·son	MONDAY lun·lun·mon	TUESDAY mar·mar·die	WEDNESDAY mer·miér·mit	THURSDAY jeu·jue·don	FRIDAY ven·vier·fre	SATURDAY sam·sáb·sam
1 May Day Maifeiertag (DE) International Workers' Day Fête du Travail (FR) Día del Trabajo (MX) National Pet Week (US)	**2** May Day (NT - AU) Labour Day (QLD - AU) Early May Bank Holiday (UK)	**3** New Moon 6:51 U.T. ●	**4** Dodenherdenking (NL)	**5** Batalla de Puebla (MX) Bevrijdingsdag (NL)	**6**	**7**
8 Mother's Day (US; AU; CAN; NZ) Fête des Mères (CAN) Moederdag (NL) Fête de la Victoire (FR)	**9**	**10** First Quarter 20:33 U.T. ◑ Día de las Madres (MX)	**11**	**12**	**13**	**14**
15 Día del Maestro (MX)	**16**	**17** Full Moon 11:09 U.T. ○	**18**	**19**	**20**	**21**
22	**23** Victoria Day (CAN) La Journée nationale des patriotes (QC - CAN)	**24** Last Quarter 18:52 U.T. ◑	**25**	**26**	**27**	**28**
29 Fête des Mères (FR)	**30** Memorial Day (US) Spring Bank Holiday (UK)	**31**	For more calendar events, visit www.browntrout.com			

APRIL 2011						
27	28	29	30	31	**1**	**2**
3	4	5	6	7	8	9
10	11	12	13	14	15	16
17	18	19	20	21	22	23
24	25	26	27	28	29	30
1	2	3	4	5	6	7

JUNE 2011						
29	30	31	**1**	**2**	**3**	**4**
5	6	7	8	9	10	11
12	13	14	15	16	17	18
19	20	21	22	23	24	25
26	27	28	29	30	1	2
3	4	5	6	7	8	9

JUNE 2011
JUIN · JUNIO · JUNI

SUNDAY dim·dom·son	MONDAY lun·lun·mon	TUESDAY mar·mar·die	WEDNESDAY mer·miér·mit	THURSDAY jeu·jue·don	FRIDAY ven·vier·fre	SATURDAY sam·sáb·sam
			1 New Moon 21:03 U.T. Solar Eclipse (Partial) 21:17 U.T.	**2**	**3**	**4**
5 Ascension Sunday Ascensión Himmelfahrt Hemelvaart	**6** Queen's Birthday (NZ) Foundation Day (WA - AU) Bank Holiday (IR)	**7**	**8**	**9** First Quarter 2:11 U.T.	**10**	**11** Queen's Official Birthday (tentative) (UK)
12 Pentecost (Whitsun) Pentecôte Pentecostés Pfingstsonntag Pinksteren	**13** Pentecost (Whit) Monday Lundi de Pentecôte Lunes de Pentecostés Pfingstmontag Queen's Birthday (AU except WA)	**14** Flag Day (US)	**15** Full Moon 20:14 U.T. Lunar Eclipse (Total) 20:12 U.T.	**16**	**17**	**18**
19 Father's Day (US; CAN; UK) Fête des Pères (CAN; FR) Dia del Padre (MX) Vaderdag (NL)	**20**	**21** Summer Solstice 17:16 U.T. National Aboriginal Day (CAN) Journée internationale des populations autochtones (CAN)	**22**	**23** Last Quarter 11:48 U.T. Fête nationale du Luxembourg (LU) Fronleichnam (DE)	**24** Fête nationale du Québec Quebec National Day Saint-Jean Baptiste (QC - CAN)	**25**
26	**27** Discovery Day (NL - CAN)	**28**	**29**	**30**	For more calendar events, visit www.browntrout.com	

MAY 2011
1	2	3	4	5	6	7
8	9	10	11	12	13	14
15	16	17	18	19	20	21
22	23	24	25	26	27	28
29	30	31	1	2	3	4
5	6	7	8	9	10	11

JULY 2011
26	27	28	29	30	1	2
3	4	5	6	7	8	9
10	11	12	13	14	15	16
17	18	19	20	21	22	23
24	25	26	27	28	29	30
31	1	2	3	4	5	6

JULY 2011

JUILLET · JULIO · JULI

SUNDAY dim·dom·son	MONDAY lun·lun·mon	TUESDAY mar·mar·die	WEDNESDAY mer·miér·mit	THURSDAY jeu·jue·don	FRIDAY ven·vier·fre	SATURDAY sam·sáb·sam
26					**1** New Moon 8:54 U.T. ● Solar Eclipse (Partial) 8:39 U.T. Canada Day (CAN) Fête du Canada (CAN)	**2**
3	**4** Aphelion 15:00 U.T. Independence Day (US)	**5**	**6**	**7**	**8** First Quarter 6:29 U.T. ◑	**9**
10	**11** Feest van de Vlaamse Gemeenschap (BE)	**12** Public Holiday (NIR)	**13**	**14** Fête nationale de la France (FR)	**15** Full Moon 6:40 U.T. ○	**16**
17	**18**	**19**	**20**	**21** Nationale feestdag (BE) Fête nationale de la Belgique (BE)	**22**	**23** Last Quarter 5:02 U.T. ◐
24	**25**	**26**	**27**	**28**	**29**	**30** New Moon 18:40 U.T. ●
31 Ramadan begins at sundown	For more calendar events, visit www.browntrout.com					

AUGUST 2011
AOÛT · AGOSTO · AUGUST

The Sun's surface gas can shoot over 300,000
miles high and span 30 Earths

SUNDAY dim·dom·son	MONDAY lun·lun·mon	TUESDAY mar·mar·die	WEDNESDAY mer·miér·mit	THURSDAY jeu·jue·don	FRIDAY ven·vier·fre	SATURDAY sam·sáb·sam
	1 Civic Holiday (CAN) Congé civique (CAN) Picnic Day (NT - AU) Bank Holiday (IR; SCT)	**2**	**3**	**4**	**5**	**6** First Quarter 11:08 U.T. ◑
7	**8**	**9**	**10**	**11**	**12**	**13** Full Moon 18:57 U.T. ○
14	**15** Assumption Assomption Asunción de María Mariä Himmelfahrt Discovery Day (YT - CAN)	**16**	**17**	**18**	**19**	**20**
21 Last Quarter 21:54 U.T. ◐	**22**	**23**	**24**	**25**	**26**	**27**
28	**29** New Moon 3:04 U.T. ● Eid-al-Fitr begins at sundown Summer Bank Holiday (ENG, WAL, NIR)	**30**	**31**	For more calendar events, visit www.browntrout.com		

JULY 2011

26	27	28	29	30	1	2
3	4	5	6	7	8	9
10	11	12	13	14	15	16
17	18	19	20	21	22	23
24	25	26	27	28	29	30
31	1	2	3	4	5	6

SEPTEMBER 2011

28	29	30	31	1	2	3
4	5	6	7	8	9	10
11	12	13	4	15	16	17
18	19	20	21	22	23	24
25	26	27	28	29	30	1
2	3	4	5	6	7	8

Astronomy

SEPTEMBER 2011

SEPTEMBRE · SEPTIEMBRE · SEPTEMBER

SUNDAY dim·dom·son	MONDAY lun·lun·mon	TUESDAY mar·mar·die	WEDNESDAY mer·mier·mit	THURSDAY jeu·jue·don	FRIDAY ven·vier·fre	SATURDAY sam·sáb·sam
				1	**2**	**3**
4 First Quarter 17:39 U.T. ◐ Father's Day (AU; NZ)	**5** Labor Day (US) Labour Day (CAN) Fête du Travail (CAN)	**6**	**7**	**8**	**9**	**10**
11 9/11 Remembrance	**12** Full Moon 9:27 U.T. ○	**13**	**14**	**15** Noche del Grito (MX)	**16** Día de la Independencia (MX)	**17**
18	**19**	**20** Last Quarter 13:39 U.T. ◑	**21** UN International Day of Peace	**22**	**23** Autumnal Equinox 9:04 U.T.	**24**
25 Daylight Saving Time begins (NZ)	**26**	**27** New Moon 11:09 U.T. ● Fête de la Communauté française (BE)	**28** Rosh Hashanah begins at sundown	**29**	**30** For more calendar events, visit www.browntrout.com	

AUGUST 2011						
31	1	2	3	4	5	6
7	8	9	10	11	12	13
14	15	16	17	18	19	20
21	22	23	24	25	26	27
28	29	30	31	1	2	3
4	5	6	7	8	9	10

OCTOBER 2011						
25	26	27	28	29	30	1
2	3	4	5	6	7	8
9	10	11	12	13	14	15
16	17	18	19	20	21	22
23	24	25	26	27	28	29
30	31	1	2	3	4	5

Astronomy

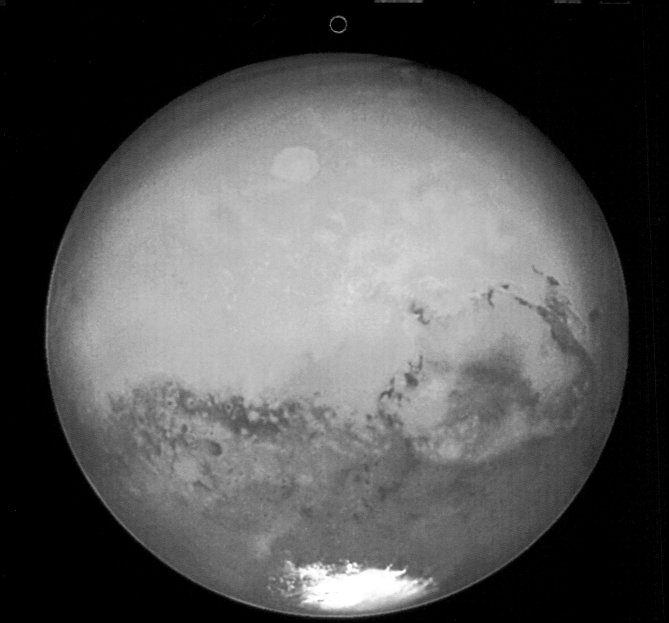

OCTOBER 2011
OCTOBRE · OCTUBRE · OKTOBER

SUNDAY dim·dom·son	MONDAY lun·lun·mon	TUESDAY mar·mar·die	WEDNESDAY mer·miér·mit	THURSDAY jeu·jue·don	FRIDAY ven·vier·fre	SATURDAY sam·sáb·sam
						1
2	**3** Labour Day (NSW, SA - AU) Queen's Birthday (WA - AU) Deutschen Einheit (DE)	**4** First Quarter 3:15 U.T. ◑ World Animal Day	**5**	**6**	**7** Yom Kippur begins at sundown	**8**
9	**10** Columbus Day (US) Thanksgiving Day (CAN) Action de grâce (CAN)	**11**	**12** Full Moon 2:06 U.T. ○ Día de la Raza (MX)	**13**	**14**	**15**
16	**17**	**18**	**19**	**20** Last Quarter 3:30 U.T. ◑	**21**	**22**
23	**24** Labour Day (NZ) United Nations Day	**25**	**26** New Moon 19:56 U.T. ●	**27**	**28**	**29**
30 European Union Daylight Saving Time ends	**31** Halloween Bank Holiday (IR) Reformationstag (DE)	For more calendar events, visit www.browntrout.com				

SEPTEMBER 2011						
28	29	30	31	1	2	3
4	5	6	7	8	9	10
11	12	13	14	15	16	17
18	19	20	21	22	23	24
25	26	27	28	29	30	1
2	3	4	5	6	7	8

NOVEMBER 2011						
30	31	1	2	3	4	5
6	7	8	9	10	11	12
13	14	15	16	17	18	19
20	21	22	23	24	25	26
27	28	29	30	1	2	3
4	5	6	7	8	9	10

Astronomy

NOVEMBER 2011

NOVEMBRE · NOVIEMBRE · NOVEMBER

SUNDAY dim·dom·son	MONDAY lun·lun·mon	TUESDAY mar·mar·die	WEDNESDAY mer·miér·mit	THURSDAY jeu·jue·don	FRIDAY ven·vier·fre	SATURDAY sam·sáb·sam
30	31	**1** All Saints' Day Toussaint Día de Todos los Santos Allerheiligen Melbourne Cup (AU)	**2** First Quarter 16:38 U.T. ◗ All Souls' Day Día de los Muertos (MX)	**3**	**4**	**5** Eid al-Adha begins at sundown Bonfire Night (UK)
6 Daylight Saving Time ends (US; CAN)	**7**	**8** Election Day (US)	**9**	**10** Full Moon 20:16 U.T. ○	**11** Veterans' Day (US) Remembrance Day (AU; CAN) Armistice (FR) Wapenstilstandag (BE)	**12**
13 Remembrance Sunday (UK)	**14**	**15** Festtag des Königs (BE) Fête du Roi (BE)	**16**	**17**	**18** Last Quarter 15:09 U.T. ◗	**19**
20 Día de la Revolución Mexicana (MX)	**21**	**22**	**23**	**24** Thanksgiving Day (US)	**25** New Moon 6:10 U.T. ● Solar Eclipse (Partial) 6:21 U.T.	**26**
27 Advent Avent Adviento	**28**	**29**	**30** St. Andrew's Day (SCT)	For more calendar events, visit www.browntrout.com		

OCTOBER 2011

25	26	27	28	29	30	1
2	3	4	5	6	7	8
9	10	11	12	13	14	15
16	17	18	19	20	21	22
23	24	25	26	27	28	29
30	31	1	2	3	4	5

DECEMBER 2011

27	28	29	30	1	2	3
4	5	6	7	8	9	10
11	12	13	14	15	16	17
18	19	20	21	22	23	24
25	26	27	28	29	30	31
1	2	3	4	5	6	7

Astronomy